# I MET A
# TEACHER ONCE

## Gabriel Malca

Copyright © 2026

All rights reserved.
No part of this publication may be reproduced,
distributed, or transmitted in any form or
by any means, including photocopying, recording,
or other electronic or mechanical methods,
without the prior written permission of the author,
except in the case of brief quotations embodied in
critical reviews and certain other non-commercial
uses permitted by copyright law.

For permission requests,
write to the author at their email address:

Author: Gabriel Malca
Email: gabriel@malca.me

ISBN: 979-8-90243-559-4

For further information about the author or their
publications,
please contact the author via email.

First Edition: 2026

A series of quiet reflections on conditioning,
pressure, and letting go.

For Setareh,
my greatest teacher.

# Contents

**I met a teacher once.**
*It happened fast.*

ast enough that my body reacted before I had time to think about what was happening. Everything in me tightened at once, like something had surfaced before I was ready to deal with it.

I could feel my face change immediately. Heat rushed upward before I could stop it. My skin felt exposed, as if it had given something away. I didn't know how much of it showed, only that it felt obvious.

I remember trying to keep my face neutral, trying not to move too quickly in either direction. I wasn't sure what neutral looked like, only that whatever was happening inside me couldn't be allowed to show.

My thoughts came in short bursts.

Did that just happen.
How bad was it.
How much did they see.

Each question arrived before the last one finished.

I replayed the moment immediately, not from the beginning, but from the point where something slipped. Where a different version of me might have been visible. I kept thinking about what I should have done instead, even though there was no longer any opening to correct it.

The feeling didn't peak and disappear. It stayed close, as if it were waiting to see how I would handle what had been revealed.

I became aware of my body in pieces. My face. My hands. My posture. Where my eyes were. I adjusted without deciding to.

When I spoke later, I could hear myself choosing words more carefully than usual. Not to say something better, but to avoid saying anything that might connect back to the earlier moment. I noticed how much effort went into sounding normal, as if normal were something I needed to regain.

I started explaining things that didn't need explaining. Adding details before anyone asked for them. Clarifying points that weren't unclear. It felt important to make sure nothing else could be misread.

At the same time, part of me wanted to do the opposite. To show steadiness. To prove competence. To make it clear that whatever had surfaced didn't define me.

Those impulses pulled in opposite directions.

Say less.
Say more.
Withdraw.
Perform.

Neither felt safe.

I became more aware of how closely I was watching myself. Not just what I said, but how I said it. How long I paused. Where I looked. Whether my face had cooled enough to pass as unchanged.

Nothing confirmed whether I had succeeded. Nothing corrected me either.

I noticed myself scanning for chances to correct the impression. To do something clean enough that the earlier moment would fade. At the same time, I tried not to stand out, in case attention made things harder to contain.

I couldn't tell which version of myself would be safer.

Over time, the specifics of the moment blurred. What remained was the feeling. The awareness that something private had appeared before I was ready.

I couldn't always remember exactly what I was reacting to anymore. I just knew that something about how I moved now required supervision.

Sometimes, when I was alone, the moment returned. Not just then, but later. Years later. The same tightening. The same sense that something about me had been exposed without my consent.

I imagined how it might have looked from the outside. How it might be remembered. How it might become a story that didn't include my explanation.

That imagined version carried more weight than what had actually happened.

Over time, I adjusted without deciding to. I paid attention differently. I measured responses more carefully. I prepared more than I used to, not because I doubted myself, but because being seen incorrectly now felt dangerous.

I noticed how often I checked myself before speaking. How often I softened reactions. How often I wondered whether something needed to be said at all.

Sometimes I compensated by being more agreeable. Sometimes by being more precise. Sometimes by trying to be better than necessary. None of it felt deliberate. It felt corrective.

Nothing else came of it. No one brought it up again.

But it stayed.

Not as something I could describe clearly, but as something that changed how much room I allowed myself afterward.

I didn't stop speaking.

I didn't withdraw completely.

I learned how to stay present without letting too much of myself show.

**I met a teacher once.**
*It didn't arrive loudly.*

It arrived as an option. One that appeared in small moments, usually right after I decided it was better not to say something. The first time felt practical. The next time felt familiar.

After embarrassment, silence felt safer. Not absence, but control. If nothing was said, nothing could surface unexpectedly. Nothing could slip out before I was ready.

There were things I noticed and didn't comment on. Thoughts that formed clearly and then stopped short of becoming words. It didn't feel like withholding at first. It felt like choosing the cleaner path, the one with fewer consequences.

Nothing bad happened when I stayed quiet.

That seemed important.

It felt like proof that silence worked.

The moment passed. The conversation moved on. Other people filled the space. I told myself that meant my silence had been the right call. That if something mattered enough, it would come back around.

It rarely did.

Silence started to feel like a way of staying out of trouble. Out of disagreement. Out of view. It reduced the number of things that could be misunderstood or challenged. It asked less of me.

When I didn't speak, I didn't have to explain myself. I didn't have to defend a thought or adjust it mid-sentence. I didn't have to deal with the feeling of saying something and immediately wishing I had said it differently.

Quiet felt efficient.

I noticed how easily I convinced myself that nothing needed to be said. That whatever I had noticed was probably obvious. Silence made those assumptions feel reasonable.

Sometimes I stayed quiet because I didn't want to be wrong. Sometimes because I didn't want to be right in a way that made things awkward. Sometimes because it felt easier to let the moment pass than to step into it.

The reasons changed. The behavior stayed.

I didn't need to decide each time. Silence started arriving before I weighed the options. By the time I noticed it, the moment had already moved on.

I began to feel the gap between what I thought and what I said. At first it was small. A word left out. A thought softened. A question held back. None of it felt like loss.

Silence gave me time. Or at least it felt that way. Time to think things through. Time to see how others responded. Time to decide whether it was worth stepping in.

Often, by the time I decided, the moment was gone.

I told myself I preferred listening. That I was being thoughtful. That I was choosing not to add noise. Those explanations held well enough to stop me from questioning the pattern.

The longer I stayed quiet, the less interruption I caused. Conversations adjusted around me. My absence from them stopped being noticeable.

That felt like confirmation.

There were moments when I wanted to speak and didn't. Not because I was unsure, but because it felt too late. Silence had already done its work.

Breaking it felt harder than maintaining it.

I noticed how silence changed what I was invited into. Fewer questions came my way. Fewer assumptions were made about what I might think. That reduction felt calm at first.

It also felt empty.

The quiet left more space than I knew what to do with.

When I did speak, it felt heavier. Like something I had been carrying and was now placing down all at once. I became aware of how much I was asking the moment to hold.

I heard it in my voice. It sounded unfamiliar, slightly out of place, like something that hadn't been used enough to settle naturally. The words landed heavier than I expected, not because they were important, but because I wasn't used to hearing myself take up that much space.

That awareness made me quieter again.

Silence taught me how to disappear without leaving. How to stay present without being involved. How to be there without being seen as part of what was happening.

It felt safe. It felt controlled.

I stopped interrupting. I stopped disagreeing. I stopped adding details that complicated things. None of that felt like a sacrifice. It felt like restraint.

There were times when silence felt like maturity. Like choosing my battles. Like staying above the need to react to everything.

Other times, it felt like relief.

What I didn't notice at first was how silence followed me after the moment passed. How it lingered even when there was nothing to avoid. How it began showing up before I had decided anything at all.

Thoughts formed and dissolved without testing whether they were worth sharing. Questions came and went. Reactions stayed internal.

I became easier to overlook. That didn't always bother me.

There were situations where silence felt like the only option. Speaking felt risky. Staying quiet felt neutral. The difference mattered.

Silence didn't ask much of me. It only required that I stay out of the way.

Over time, it became automatic. Quiet arrived first. Speech became the thing that needed justification.

When something went unsaid, it was hard to tell whether it had been a choice or a habit.

I noticed how often I replayed conversations later. Not to change what I had said, but to imagine what I might have said if I had spoken at all. Those imagined versions stayed longer than the moments themselves.

Silence made it easy to leave without leaving a trace. It also made it easy to feel like I wasn't fully there.

Nothing marked the shift. There was no moment where I decided to stop speaking. No clear trade. Just a gradual narrowing of what I considered worth voicing.

Silence didn't solve anything. It reduced friction. It lowered risk. It kept things smooth.

It also stayed.

It stayed when something mattered. It stayed when I had something to add. It stayed when being quiet no longer felt neutral.

By the time I noticed, silence had already taught me how to step back before being asked.

And it didn't tell me what it was costing.

**I met a teacher once.**
*It appeared when action was possible, but commitment wasn't.*

When I could move forward, but doing so would collapse too many other paths at the same time.

Delay stepped in there.

It felt like restraint. Like patience. Like choosing not to rush something that mattered.

I told myself I was being responsible.

There were futures I wasn't ready to close yet. Versions of my life that couldn't exist together, but that I wasn't willing to let go of. Delay gave me a way to keep them alive without choosing between them.

Nothing forced a decision.

The days kept moving. Life stayed functional. I could act in small ways without committing fully. I could progress without arriving anywhere that required me to stay.

That middle space felt safer than choosing.

Delay made the present livable by postponing the cost. It allowed me to keep things intact without resolving the tension that held them together.

I didn't feel stuck.

I felt suspended.

I told myself I was waiting for the right moment. For clarity. For something external to make the decision cleaner than it was. That waiting felt thoughtful.

It also felt endless.

I noticed how delay protected me from loss. As long as I hadn't chosen, nothing was gone yet. No door had fully closed. No world had been left behind.

But nothing had been fully entered either.

I lived in that overlap longer than I realized. Long enough for waiting to become a way of being. Long enough for indecision to feel like stability.

Others moved ahead of me in different directions. I stayed where multiple paths were still possible.

That felt like freedom.

It was also a way of avoiding grief.

Delay let me care without committing to the consequences. It allowed me to stay invested while keeping an exit available.

I told myself that was wisdom.

What I didn't notice at first was how much of my life I was spending in anticipation rather than presence. How often I treated the current moment as something to endure rather than inhabit.

I was always preparing for a future that hadn't agreed to arrive.

Eventually, delay began to shape me. I became someone who could hold a lot without choosing. Someone who could wait indefinitely without admitting that waiting was a decision.

The moment stayed open.

Not because it needed more time.

But because closing it would have meant becoming someone I wasn't ready to be yet.

**I met a teacher once.**
*It arrived all at once.*

There was no time to prepare, no moment to ease into what was being asked. One second things felt open. The next, it felt like everything depended on what happened next.

My thoughts became loud immediately. Not careful or measured. Just loud.

What if this goes wrong.

What if I can't do this.

What if this is the moment that shows I'm not as capable as I think.

There wasn't space to slow them down. Each thought pulled another one in behind it, faster than I could sort through them. Consequences appeared before anything had actually happened.

The pressure didn't come from the task itself. It came from what it would mean if I made the wrong choice.

Making the wrong choice didn't feel contained. It felt like it would spread outward, touching things it didn't belong to, changing how past effort would be remembered.

There wasn't a small version of the outcome. Either this worked, or it meant something was wrong.

That was the pressure.

It filled the space immediately. There was suddenly less room for anything else. Less room for doubt. Less room for choice. Less room to pretend this didn't matter as much as it clearly did.

What made it worse was how familiar it felt. I had been in moments like this before. Different setting. Same feeling.

My mind jumped ahead without asking permission.

If this goes wrong, then that changes.

If that changes, then something else follows.

The chain kept going, even though none of it had happened yet.

Everything collapsed into the present. What had gone well before didn't help. What might go well later didn't either. Only what happened next seemed to matter.

I could feel the expectation, even if no one said it directly. It was there in the attention, in the pause before something had to happen. It felt like a test I hadn't agreed to take.

The pressure made it seem like there was a right way to be in that moment, and that I had to find it immediately. Confident, but not forced. Capable, but not strained. Calm, but not detached. Anything else felt risky.

There wasn't time to try different versions. There wasn't time to hesitate.

Fear showed up as urgency. As the need to act before the worst version of things had a chance to arrive.

Every possible mistake surfaced at once. Things I might miss. Things I had little experience with. Weak spots I hoped wouldn't matter here. My mind treated them as warnings, not questions.

It was easy to imagine how this could go wrong. Much harder to imagine it simply being fine.

Doing nothing didn't feel neutral. Waiting didn't either. Even choosing carefully felt risky, because choosing carefully took time, and time felt like the one thing I didn't have.

I felt the urge to get out of it, even while knowing there wasn't anywhere to go. Leaving would mean making the wrong choice. Pausing would mean making the wrong choice. Asking for help would mean making the wrong choice. Every alternative pointed back to the same conclusion.

So I moved forward anyway.

I didn't wait for the fear to settle. I acted with it still running, listing everything that could go wrong while I tried to keep it from happening.

The pressure didn't care whether I felt ready. It didn't care whether the situation was fair. It only cared that the moment was here and that something had to happen.

What made it heavier was the sense that this wouldn't be the last time. Even while I was still inside it, I was already thinking about how often it would return. How many times I'd have to do this again if it went well. How little room there was to get it wrong even once.

It felt like whatever happened here would follow me into the next moment.

The pressure made it hard to feel human. Hard to stumble and recover. Hard to be unsure and then find footing. It made the outcome feel final, even though nothing in life actually is.

Even success didn't feel safe. If it worked, it felt like something I'd have to protect. Something I'd be expected to repeat. The thought of having to do it again sat right next to the thought of getting it wrong.

Time stretched strangely. Everything felt rushed, but each second also felt heavy. My thoughts circled the same fears, repeating when they couldn't find anything new.

What if this is where it breaks.

What if this is where it shows.

What if this is where everything falls apart.

The moment didn't wait for those questions to be answered.

Afterward, the pressure didn't disappear right away. It stayed close, replaying what had just happened, checking for signs that it had been right to be there.

It didn't need to be right to stay convincing.

What it left behind wasn't clarity. It was conditioning. A way of carrying fear while still functioning. A way of treating moments like this as inevitable instead of exceptional.

I learned how to move with the noise still running. How to act while imagining the worst. How to keep going even when the cost felt high.

It didn't explain itself.

It just waited for the next moment where too much seemed to depend on what I did.

**I met a teacher once.**
*It showed up when effort stopped feeling decisive.*

When things were going well, but not well enough to explain why. When momentum existed without guarantees. When skill had brought me this far, but couldn't promise what came next.

Something still had to give.

I felt it most when outcomes were no longer proportional to effort. When preparation reached its limit and whatever happened next felt out of my hands.

Trying something and not knowing if it would work.

Waiting on a decision that could tilt everything.

Realizing that no amount of refinement would make the answer arrive faster.

There were moments when it felt like the result had already been decided somewhere else. Not by people. Not by systems. By something harder to locate.

Call it luck.
Call it timing.
Call it the universe.

Whatever it was, it wasn't listening to me.

I could do everything right and still lose. I could prepare endlessly and still be surprised. That possibility stayed close, especially when things were going well.

Because that's when it felt most fragile.

If things improved too smoothly, I waited for the reversal. If momentum held, I questioned how long it was allowed to. I didn't trust that the arc would keep bending in my favor.

Something always had to balance out.

That belief didn't come from evidence. It came from exposure. From knowing that outcomes don't always reward effort evenly. From seeing people do everything right and still get shut out.

From knowing I could be next.

Uncertainty made success feel provisional. As if it were being loaned rather than earned. As if something unseen could reclaim it without explanation.

That's when the question shifted.

Not *what if this doesn't work.*

But *what if it means something.*

If things fell apart, was that just circumstance, or was it a verdict. If they went wrong, was that random, or corrective.

I didn't know which frightened me more.

I noticed how easily outcomes started to feel moral. How quickly failure turned into interpretation. Not about strategy, but about worth.

Was I supposed to have this.

Had I overreached.

Was this where it corrected.

Uncertainty didn't feel like not knowing what would happen.

It felt like not knowing whether I was allowed to want it at all.

I kept moving. I kept working. I kept preparing.

But underneath all of it was the sense that something else was deciding, and that whatever it chose would arrive without justification.

That was the fear.

Not that things wouldn't work.

But that if they didn't, it would feel final in a way effort couldn't undo.

**I met a teacher once.**

*It showed up when things became predictable enough to manage.*

After uncertainty, expectation felt like relief. The field had narrowed. Outcomes weren't guaranteed, but they were no longer infinite. I could see the variables. I could influence them. I could prepare.

That felt safer.

I started forming expectations about how things would go. About how people would respond. About what effort should produce if applied correctly.

I expected work to be met with work.

At work, I expected people to care the way I did. To think through problems deeply. To push harder when things mattered. I told myself this was reasonable. We were building something. We were moving fast. Commitment was part of the deal.

I paid them a salary. I paid myself in equity. I didn't treat that difference as meaningful.

I expected full engagement in return.

I expected people to stay late without being asked. To think ahead. To surprise me. To feel the same urgency I felt. When they didn't, I framed it as a gap in standards rather than a difference in incentives.

I didn't question the expectation.

I questioned their dedication.

I also had expectations of myself.

I expected to succeed. Not casually. Decisively. I believed I was smart. Smarter than most. I had ideas that worked. I could see paths others missed. I wanted that to be recognized.

I expected respect. I expected people to listen. I expected my thinking to carry weight.

At the same time, I expected myself to appear humble. To downplay confidence. To act as if recognition didn't matter, even while waiting for it to arrive.

That contradiction stayed unresolved.

I expected success to prove something. Not just that my ideas were good, but that I was worthy of the position I occupied. That the effort had been justified. That the risks had been correct.

I expected outcomes to validate me.

Expectation also became a way of organizing behavior. Mine and everyone else's. When things went well, I took it as confirmation that the system was working. When they didn't, I looked for where expectations had been violated.

Someone hadn't tried hard enough.

Something hadn't been planned carefully enough.

I hadn't prepared well enough.

Expectation made the world feel legible. It reduced chaos. It replaced uncertainty with standards. It gave me a framework for interpreting outcomes.

It also narrowed my tolerance.

When people didn't meet expectations, I felt let down. When results fell short, I felt misled. When effort didn't produce what it was supposed to, I took it personally.

Expectation made disappointment feel justified.

I kept refining it. Adjusting assumptions. Tightening standards. Rehearsing scenarios. Preparing responses in advance.

The more I expected, the more I believed things should make sense.

And when they didn't, I assumed something had gone wrong.

Either in others.

Or in me.

**I met a teacher once.**
*In my community, being known was the goal.*

People wanted to be referenced. To be named. To be pointed to as proof that something worked. Success was meant to travel. It was meant to be visible.

I didn't want that.

Not because I rejected it, but because I didn't trust it. Being seen felt dangerous. Like inviting attention that wouldn't leave once it arrived.

I believed in the evil eye.

Not abstractly. Practically. If something was going well, it felt fragile. If it was visible, it felt at risk. I didn't believe I was entitled to keep what I had.

Anything could be taken back.

That belief shaped how I moved. I stayed careful. I avoided the spotlight. I told myself this was humility. That wanting less attention meant I was grounded.

It wasn't humility.

It was fear.

Once you're visible, people form expectations. Not privately. Collectively. A version of you starts circulating, and you no longer control how it's used.

Being successful meant being available. For advice. For support. For introductions. For charity. For ideas that needed backing.

Saying yes felt required.

Saying no felt like exposure of a different kind.

Because no one just hears no. They hear refusal. Disappointment. A story about who you really are once the image wears thin.

You didn't help.

You didn't show up.

You didn't participate in something that mattered.

That version travels too.

Visibility doesn't ask whether you're capable of carrying it. It just keeps producing demands. Every interaction becomes proof. Every absence becomes evidence.

I felt the pressure to behave correctly. To be generous without limits. To be patient without boundaries. To never let frustration show.

Good people don't get tired.

Good people don't say no.

Good people don't protect themselves at the expense of the group.

Exposure made those rules feel constant.

I watched how quickly people talked. How often names came up. How easily stories formed. Once you were known, you were never discussed in just one way.

There was always a version.

Any time my name was mentioned, I imagined it being weighed. Compared. Interpreted. Not against facts, but against expectations.

Would the story be good.

Would it be flattering.

Would it be real.

I learned how to manage that risk. I softened edges. I avoided conflict. I chose language that couldn't be misread. I stayed agreeable longer than I should have.

Not because I wanted to be liked.

Because being disliked felt costly.

Exposure didn't threaten my worth.

It threatened my freedom.

Once people were watching, everything felt like it carried consequences. Even private decisions felt public. Even silence felt noticeable.

I wasn't hiding.

I was managing.

I could still succeed. Still build. Still move forward. The difference was how much of myself I allowed to show while doing it.

Over time, fewer things felt safe to release into a space where they couldn't be contained.

Being known didn't feel like recognition.

It felt like surveillance.

And I learned how to stay visible without ever fully stepping into the light.

**I met a teacher once.**
*It wasn't loud or sharp.*

It showed up as a quiet sense that being okay was slightly inappropriate. That comfort required an explanation I didn't have.

When something went well for me, I didn't settle into it. I noticed it, then stepped back. Success felt exposed, like it needed to be handled carefully so it wouldn't say the wrong thing about me.

I learned to soften good news.

There were moments when I felt ahead of others. More stable. More comfortable. Those moments didn't feel clean. They came with an urge to downplay, to qualify, to make sure no one thought I believed I deserved it.

I rarely stayed with pride for long.

I compared my position to people close to me. If someone struggled while I was doing well, the difference felt heavy. Not because I had caused it, but because I was visible inside it.

That visibility made ease feel questionable.

At work, there were times I benefited from systems I didn't fully believe in. I was rewarded for things that didn't line up cleanly with how I wanted to see myself. The reward still arrived. That unsettled me.

I told myself to be grateful. Gratitude didn't settle it.

I became careful with enjoyment. Careful about celebrating wins. Careful about letting things feel good without qualification. It felt safer to stay slightly restrained than to risk appearing untroubled.

I didn't want to look like someone who wasn't paying attention.

I noticed how quickly I tried to compensate. Giving more. Explaining more. Being useful in ways that felt corrective rather than free.

Those gestures felt necessary.

When people praised me, I redirected. When things went smoothly, I looked for where they might not have. When outcomes favored me, I waited for the part where they wouldn't.

That waiting became familiar.

There were times when I felt responsible for things that weren't mine to carry. Not because I had done anything wrong, but because being unaffected felt wrong on its own.

I didn't need blame to feel implicated.

I learned how to stay slightly behind my own life. Not fully refusing it, but not stepping into it either. I occupied space cautiously, as if it were borrowed.

That caution felt ethical.

It also dulled things.

Rest came with the need to look occupied.

Ease felt like something that had to be justified.

When I enjoyed something, part of me stayed alert, ready to pull back if it went too far.

Over time, this posture spread. Success didn't land fully. Relief didn't settle. Good moments arrived, then passed through quickly, as if lingering would be a mistake.

I didn't feel bad.

I felt constrained.

Constrained in what I allowed myself to enjoy. In how much ease I trusted. In how openly I inhabited what was working.

It wasn't that I believed I was wrong.

It was that being unburdened felt irresponsible.

So I stayed careful.

Guilt didn't demand punishment. It demanded moderation. Restraint. A constant awareness of balance.

I learned how to live as if comfort needed supervision.

And I never felt fully permitted to stop supervising it.

**I met a teacher once.**

*It showed up when things stopped growing.*

The effort stayed the same. The attention stayed the same. The outcome settled into a range that no longer responded the way it used to.

I noticed it quickly.

At first, I treated it as resistance. Something to push through. I added more effort. Took on more responsibility. Stretched past what felt reasonable. That approach had worked before.

This time, the boundary held.

The limit wasn't dramatic. It didn't announce itself. It simply stayed where it was. Progress slowed, then flattened, as if the space ahead had narrowed without explanation.

I could sense what existed beyond it. A different scale of life. More room. More ease. A version of success that required expansion to stay intact.

I wanted that version.

The wanting stayed active even as movement stalled. I reached while conditions stayed unchanged. The tension between desire and permission remained.

I kept testing the edge.

Spending adjusted before I chose it to. Options narrowed. Not because I preferred less, but because the numbers required it. The rules shifted quietly. The range tightened without asking for agreement.

I pushed against that.

There was frustration in being held where I was. In seeing paths that were still technically possible but practically blocked. In knowing that a small shift would change everything while having no way to produce it.

The limit pressed back.

Fear entered gradually. Fear of slipping backward. Fear of misjudging what the situation could support. Fear of becoming dependent on a version of life that couldn't sustain itself. That fear didn't slow me down.

It made me more controlled.

I tightened spending. Delayed decisions. Reconsidered upgrades that once felt inevitable. Each adjustment felt like retreat, even when framed as prudence.

I didn't want to shrink.

I wanted to advance.

The restraint felt enforced rather than chosen. I followed the rules because the consequences of breaking them were clear. Compliance became practical.

I stayed inside the range because stepping outside it carried risk I wasn't willing to take.

At the same time, I kept imagining what it would demand if the ceiling lifted again. The pace. The pressure. The obligation to keep expanding just to hold the position. That awareness didn't soften the desire.

It complicated it.

I didn't feel grateful for what I had. I stayed aware of what remained out of reach. The gap stayed present.

I tried to make sense of the limit. To frame it as timing. As discipline. As something that served a longer arc. Those explanations helped me continue.

They didn't move the boundary.

The ceiling stayed where it was. I adjusted how I moved beneath it. Planned more carefully. Measured risk differently. Learned which edges I could press without breaking what I had.

I kept reaching.

Some days, the restraint felt responsible. Other days, it felt confining. I couldn't always tell which one I was living.

The limit remained.

And I carried that awareness forward, not as understanding, but as something that had to be accounted for every time I reached beyond what was already allowed.

**I met a teacher once.**
*It grew slowly.*

Through staying. Through continuing to do what was required long after it stopped feeling chosen. I kept showing up, kept complying, kept holding things together. The absence of rupture allowed it to build.

Each day asked for the same effort. The same restraint. The same patience. I met those requests without protest. On the surface, everything held.

Inside, something tightened.

I noticed it first in small reactions. A delay before responding. A hesitation where generosity had once come easily. I still did what was expected, but the cost increased.

The effort felt heavier.

I told myself this was endurance. That staying showed strength. That commitment meant absorbing discomfort without complaint. Those ideas kept me in place.

The tension remained.

Resentment took shape in what I began to ration. I became precise about what I gave and when. Care became measured. Energy was allocated. I kept track without deciding to.

Everything stayed unspoken.

I complied while feeling separate from what I was doing. The actions continued. The connection thinned. I stayed present, but generosity receded. I remained available, but openness narrowed.

That distance held.

I noticed how easily irritation surfaced. Not explosive, but sharp enough to register. Small requests carried more weight. Interruptions felt personal. Expectations pressed simply by existing.

Anger didn't arrive loudly.

A sense of being used settled in instead. Not only by people, but by the situation I had agreed to remain inside.

Resentment carried memory. It remembered how long I had stayed quiet. How often I had adjusted. How many times I had swallowed frustration to keep things moving.

I began conserving myself. Doing what was required to maintain balance. Avoiding the extra effort that once felt natural. I stayed close enough to function.

Staying took work.

I noticed how patience became performative. I acted calm while counting the cost. I listened while carrying the weight of what had already been absorbed.

The experience felt compromised.

Resentment didn't move me toward leaving. It made leaving feel complex. Investment pulled against exit. Time spent resisted being written off.

So I stayed.

Over time, movement slowed. Initiative softened. Curiosity narrowed. Care continued, but with caution. I protected myself from giving more than I could afford to lose.

That protection felt necessary.

I became efficient. Reliable. Predictable. Those traits kept things running. They also preserved the existing shape.

The days blended together. Dense rather than unpleasant. Each one carrying a trace of the last.

Resentment grew through accumulation. Through effort that no longer matched what returned. Through imbalance that stayed visible.

I carried awareness of that gap.

There were moments when I imagined stepping out of it. Pausing. Resetting. Those thoughts arrived quietly and passed the same way. Disruption felt expensive.

Staying felt easier than beginning again.

The situation stayed intact. Stability held. Change never forced its way in. From the outside, it resembled persistence.

From the inside, it felt like carrying something that was never put down.

Resentment didn't ask for release. It asked for endurance. It taught me how to keep moving without momentum. How to continue while holding back.

I learned how to live inside that tension.

And it stayed, shaping what I was willing to give long before anything finally asked me to stop.

**I met a teacher once.**
*It didn't show up when things broke.*

It showed up when they didn't. When nothing collapsed, but nothing improved either. When the situation stayed intact long enough for staying to feel like the responsible choice.

I was already committed by the time doubt arrived. Time had been spent. Effort had been given. Walking away would have meant writing all of that off at once.

So I stayed.

At first, staying felt patient. Like endurance. Like the kind of persistence people admire when it eventually pays off. There was always a reason to keep going. Something unfinished. Something close enough to working to justify more time.

Progress didn't stop. It slowed. Just enough movement to make leaving feel premature. Just enough promise to keep hope intact.

I adjusted my expectations without noticing. I told myself outcomes take time. That things settle late. That this stage was necessary before anything solid could arrive.

Waiting felt disciplined.

The longer I stayed, the harder it became to leave. Not because things improved, but because staying accumulated weight. Time invested became something I felt responsible for protecting.

I didn't want it to be for nothing.

There were moments when I noticed the mismatch. When effort and return no longer aligned. Those moments didn't last. I explained them away. I reframed. I reminded myself how far I had already come.

Starting over felt worse than continuing.

I learned how to live inside that tension. How to keep functioning while something underneath stayed unresolved. I felt both stuck and committed to seeing it through.

That commitment became part of how I understood myself.

Others moved on from similar situations. Changed direction. Took exits when they appeared. I told myself my patience was different. That I was willing to stay longer for something that mattered.

I didn't see how much staying was shaping me.

Time passed. The situation stayed roughly the same. I made small adjustments to keep it workable. I increased pressure to resolve it while asking myself when it would change.

I kept going.

Failure didn't announce itself. It accumulated. It took shape slowly, through repetition. Through explaining the same thing again. Through waiting for the same shift that never arrived.

When the end finally came, it didn't feel dramatic.

It felt overdue.

What hurt wasn't that it hadn't worked. It was how long I had stayed after it stopped moving. How much of myself I had continued to give to something that wasn't going to meet me where I was.

I looked back and saw how gradually my standards had adjusted. How I accepted less to justify staying. How staying out of patience became staying out of habit.

Leaving didn't feel like relief.

It felt like acknowledging how long I had already known.

Failure wasn't the ending.

Failure was the stretch of time before it, where I kept going because stopping would have meant admitting it wasn't going to change.

That stretch stayed with me.

It showed up later, whenever I hesitated to walk away. Whenever I waited longer than I needed to. Whenever I confused endurance with loyalty.

I learned how easy it is to stay past the point where something deserves you.

And how hard it is to admit when that point has already passed.

**I met a teacher once.**
*Someone else's progress entered the room.*

Nothing about my situation had changed. The work was the same. The effort was the same. The outcome hadn't moved.

It just sat next to theirs.

What I had been satisfied with a moment earlier no longer held the same weight. It didn't disappear. It thinned. The sense of having arrived anywhere loosened.

I noticed how quickly my attention shifted. Not away from my own work entirely, but back and forth. From what I had done, to what they had done. From my position, to theirs.

I looked at what they had built. Then I looked back at what I had. The comparison didn't ask whether the paths were similar. It didn't ask how long each thing had taken. It didn't care what had been sacrificed along the way.

It only asked who was further along.

I could still name progress. I could still point to movement. None of it landed the same way once it existed beside someone else's result.

What had felt earned began to feel partial. What had felt solid began to feel dependent on context.

I found myself replaying the timeline. Where I had slowed down. Where I had chosen differently. Where momentum might have been lost. The questions came without invitation.

I didn't answer them.

They stayed.

They showed up again when I tried to focus. When I tried to enjoy what I had already built. When I tried to measure things on my own terms.

Enjoyment shortened. Satisfaction became conditional. Before letting myself feel settled, I checked where I stood.

There was always another reference point.

Even when I met one standard, another appeared. Even when I closed one gap, another became visible. The sense of being "on track" kept moving just far enough ahead of me.

I noticed how quickly my position started to feel like something that required context. As if it might be misunderstood if left on its own. As if the gap needed explaining.

No one asked for that explanation.

The pressure came from inside.

Goals adjusted quietly. Not because they no longer mattered, but because their shape had changed in comparison. What counted shifted. What had once felt substantial now felt provisional.

I raised the bar without naming it. I postponed satisfaction without deciding to. I treated where I was as temporary, even when it had once felt like an arrival.

I still moved forward.

I just stopped letting where I was be enough on its own.

**I met a teacher once.**
*It arrived as space.*

Not emotional space, but physical distance. A shift that placed me outside the
field where I had been continuously
seen. Outside the patterns that had shaped how I
moved, how I spoke, how I understood myself.

The change was immediate.

Without familiar eyes on me, something
loosened. I felt less watched. Less interpreted.
Less required to explain who I was becoming in
real time.

I moved through my days without having to
account for them.

That absence felt unfamiliar at first. I had
grown used to being referenced. To knowing
where I stood relative to others and what that
position implied.

Distance removed that orientation.

The break created room.

With fewer signals coming in, my energy stayed closer to me. I didn't have to defend it or shape it. I could spend time without needing to justify how it looked.

I noticed how much effort had once gone into managing proximity. Staying close enough to belong. Far enough to breathe. Distance simplified that tension.

Some relationships faded quietly. Not through conflict or decision. They lost momentum once constant contact was no longer holding them in place.

Others changed shape. Conversations felt different when they weren't embedded in shared expectations.

Distance made it possible to meet people as they were, rather than as they had been.

Being out of sight reduced the need to perform. Without an audience, certain behaviors no longer served a purpose. Some habits softened. Some concerns lost urgency. I didn't feel the same pressure to stay busy or prove relevance.

Time opened up.

I could explore myself without narration. Without turning experience into something that needed to be legible to others. Curiosity became quieter and more direct.

I became aware of how much identity had once been reinforced through repetition. Seeing the same people. Hearing the same language. Responding to the same expectations. Distance interrupted that rhythm.

In its place came observation.

I watched how communities functioned from the outside. How belonging shaped behavior. How consensus hardened into certainty. How people took on shared scripts without questioning them.

Those patterns became easier to see from a distance.

I noticed how often people moved together, guided by the same assumptions. How disagreement quickly became uncomfortable. How loyalty sometimes mattered more than honesty.

I understood why that was appealing.

Belonging relieved people from having to decide everything on their own. It offered protection, certainty, direction. It also asked for consistency.

Being away made that tradeoff clearer.

Without constant proximity, those expectations lost urgency. Opinions softened. Positions relaxed. I could accept people as they were, instead of measuring them against what they represented.

When I returned, even briefly, familiar faces felt different. They carried less weight. They didn't stand in for anything larger.

I no longer needed them to confirm anything about me.

Community changed shape once it stopped defining me. I could appreciate what it offered without needing it to approve my direction. I could see where it supported growth and where it narrowed it.

That clarity came from stepping back, not pushing against it.

Distance also quieted something else.

The constant hum of anticipation eased. Without immediate judgment, my body settled. I wasn't bracing as often. I wasn't preparing responses before they were needed.

Presence became simpler.

I noticed how anxiety had once thrived on proximity. On interpretation. On the sense that something was always being evaluated. Distance thinned that atmosphere.

The edge eased.

I could still care deeply. Still commit. Still work with intensity. The difference was where that intensity lived. It no longer needed to be witnessed.

I didn't have to disappear to feel this way. I didn't have to withdraw completely. Physical separation was enough to loosen the grip of expectation.

It showed me how much of my life had been shaped by closeness.

Distance wasn't avoidance.

It was space that allowed things to settle.

A place where energy could return to itself. Where relationships could adapt or fall away without force. Where identity could shift without announcement.

I stayed there long enough to feel the difference.

Long enough to know which parts of myself needed space to exist.

Long enough to see that clarity often arrives after separation, not during engagement.

Distance didn't tell me who to become.

It quieted what had been telling me who to be.

And that space stayed with me, even when proximity returned.

**I met a teacher once.**
*It arrived after distance had already done its work.*

Not as a moment, but as a gradual thinning. Things that once felt central no longer held the same grip. Not because I pushed them away, but because I no longer needed to hold them in place.

I noticed it first in what stopped asking for effort. Attachments softened. Old pressures loosened. Expectations that had once felt urgent no longer demanded the same attention. I didn't feel like I was giving anything up. I felt like I was no longer carrying as much.

Letting go didn't feel dramatic.

It felt quiet, almost unremarkable.

The urgency that had shaped so many of my decisions faded without announcing its departure. I stopped maintaining certain versions of myself simply because there was no longer a reason to. What I had once protected out of habit no longer felt worth defending.

Some losses were visible. Relationships changed. Work shifted. Structures I had relied on stopped holding the same importance. But those changes didn't feel like collapse. They felt like rebalancing.

Other losses were harder to name.

I stopped identifying as strongly with roles that had once felt essential. Stories I had used to understand myself lost their authority. The need for continuity weakened. I didn't rush to replace any of it. I let the space stay open.

That openness was unfamiliar.

Not empty. Just unclaimed.

Distance had given me time to sit without reacting. Without performing. Without having to prove that I was making sense. In that stillness, earlier conditioning became easier to recognize.

Pressure no longer set the pace. Expectations lost their leverage. Uncertainty softened once it wasn't being treated as a verdict on who I was or where I was headed.

What surprised me most was how resentment began to dissolve. Not because I decided it didn't matter, but because I could finally see it clearly. Without the weight of proximity, the past rearranged itself.

Nothing had been done to me.

Everything had been done for me.

From that wider view, events no longer felt targeted or corrective. They felt necessary. The meaning I had assigned them loosened. The charge faded. What remained was understanding.

Forgiveness didn't arrive as a gesture.

It arrived as recognition.

I could forgive others because I no longer needed them to have been different. I could forgive myself because I no longer needed the past to justify who I was becoming.

Loss changed how I evaluated effort. What once demanded loyalty now felt optional. What I had maintained out of obligation no longer asked to be sustained.

I became selective without deciding to.

Some connections thinned because I no longer met them with the same urgency. Some ambitions faded because they belonged to a version of me that no longer needed to be upheld. None of it felt forced. It felt natural.

I noticed how much of my life had been organized around momentum. Around keeping things moving simply because they already were. When that momentum broke, I didn't rush to replace it.

There was relief in not needing to be consistent. In not having to explain change. In not justifying why certain things no longer mattered.

That relief came with uncertainty. With the feeling of moving without a map. Without structures that once told me who I was supposed to be.

I didn't try to resolve that.

Loss didn't make me empty.

It made me less attached.

In that loosening, something else became possible. Not certainty. Not answers. But a gentler relationship with myself. Without judgment constantly pressing in, care became easier to extend. Toward others. Toward family. Toward the person I had been while learning all of this.

I carried what remained forward, without insisting it turn back into what it had been.

**I met a teacher once.**
*For a long time, authority lived outside me.*

It came from structure. From rules. From people who spoke with certainty and carried weight. It arrived through religion, community, and shared frameworks that told me where I stood and which direction mattered.

Authority charted the path.

As long as I stayed aligned with it, movement felt justified. Effort had context. Progress could be measured. There was a ladder to climb and a way to know where I was on it.

I didn't move cautiously inside that structure. I pushed forward. I wanted to do well. I wanted to excel. Being aligned didn't remove ambition. It focused it.

Authority gave effort somewhere to land.

When things went well, it felt like confirmation. When they didn't, it felt like instruction. Experience pointed beyond itself, and I trusted that direction.

Over time, parts of that structure fell away. Not suddenly. One certainty at a time. A belief that no longer held. A role that stopped fitting. An obligation that lost its hold on me.

Each loss opened space.

At first, I assumed something else would fill it. Another framework. Another voice with clarity. I stayed attentive, waiting for direction to return in a different form.

It didn't.

The absence was quiet. There was no declaration that things had changed. Decisions still appeared. Action was still required.

What changed was where direction came from.

Without authority charting the path, decisions lost their ranking. They no longer pointed upward toward approval or alignment. They didn't need to be measured against a shared standard.

They pointed inward.

The absence didn't add weight. It removed an audience. There was less to perform for. Less to prove. Intention mattered more than outcome, because comparison had thinned.

I could still push forward. Still aim high. Still care deeply about doing things well.

The difference was who that effort belonged to.

I noticed how much energy had once gone into positioning. Into being seen in the right place, moving the right way, advancing according to rules that weren't mine.

That energy loosened.

The space it left behind felt unfamiliar. Without a ladder, ambition lost its outline. Without hierarchy, progress didn't announce itself.

I felt the pull to replace what had been lost. To find new guides. New teachers. New structures that promised direction without the rigidity of the old ones.

I recognized the pattern.

Placing someone else ahead of me had once felt stabilizing. It had also transferred authorship.

Direction came from alignment rather than discernment.

I didn't want to recreate that.

So I resisted replacing authority with a softer version of itself. I stopped borrowing certainty. I stopped looking for someone else to stand above the decision.

Choices became simpler without becoming smaller.

They didn't carry endorsement or condemnation. They carried intent. What mattered was whether I could stand behind them, not how they would be ranked later.

Outcome became information rather than judgment.

Some pursuits lost their appeal once they no longer conferred status. Others gained clarity once they belonged fully to me.

What remained was direction without hierarchy. Movement without permission. Effort without comparison.

I stayed there, choosing without needing the choice to be approved, knowing nothing would arrive later to declare it right or wrong.

**I met a teacher once again.**
*It felt familiar when it returned.*

Not identical, but close enough that my body recognized it before my thoughts did. The same tightening. The same sense that too much depended on what happened next.

The situation was different. The stakes were different. I was different too.

The pressure still arrived quickly.

The noise showed up the way it always had. The questions. The imagined consequences. The awareness of how far this moment could reach if it went wrong.

That part hadn't softened.

What had changed was how much authority I gave it.

I noticed the urge to rush. To perform. To compress myself into whatever shape felt most acceptable under the weight of expectation. That urge was still there.

I didn't follow it immediately.

I stayed with the moment long enough to feel the pressure without letting it decide for me. Long enough to notice that urgency didn't always mean accuracy.

The pressure still wanted action. It still framed hesitation as risk. It still suggested that failure here would carry consequences beyond what was visible.

I acknowledged it.

Then I did what felt possible instead of what felt demanded.

The outcome mattered. I didn't pretend it didn't. But it no longer felt final. One moment no longer stood in for everything that came after.

I felt the old reflex reach for control. For certainty. For reassurance that this would work and keep working.

I let that go.

I didn't need the moment to confirm anything about who I was.

The pressure stayed, but it no longer filled the entire space. There was room alongside it. Room to choose. Room to respond instead of react.

I didn't feel calm.

I felt steadier.

Pressure still shows up. It likely always will. But it no longer feels like something I have to obey to survive the moment.

It's something I can notice without surrendering to.

I notice it, and I take the next step anyway.

www.ingramcontent.com/pod-product-compliance
Lightning Source LLC
Chambersburg PA
CBHW020458160726
47991CB00007B/2715